spot

EXTREME WEATHER

WILDFIRES

by Anastasia Suen

AMICUS | AMICUS INK

sparks

trees

Look for these words and pictures as you read.

helicopter

firefighters

A campfire is fun.
A wildfire is not.

Most wildfires are started by people. A small fire can get big fast.

sparks

See the hot sparks?
The wind blows them.
They fly away and burn.

trees

See the trees?
Fire burns the branches.
The flames move to other trees.

305

helicopter

See the helicopter?

It drops water.

It keeps the fire from spreading.

firefighters

See the firefighters?
They dig a fire line.
They try to contain the fire.

After five days, this fire burns out.
Good job, firefighters!

sparks

trees

Did you find?

helicopter

firefighters

Spot is published by Amicus and Amicus Ink
P.O. Box 1329, Mankato, MN 56002
www.amicuspublishing.us

Library of Congress Cataloging-in-Publication Data
Names: Suen, Anastasia, author.
Title: Wildfires / by Anastasia Suen.
Description: Mankato, MN : Amicus, [2021] | Series: Extreme weather | Audience: Ages 4-7 | Audience: Grades K-1
Identifiers: LCCN 2019036316 (print) | LCCN 2019036317 (ebook) | ISBN 9781681519463 (library binding) | ISBN 9781681525938 (paperback) | ISBN 9781645490319 (pdf)
Subjects: LCSH: Wildfires—Juvenile literature.
Classification: LCC SD421.23 .S84 2021 (print) | LCC SD421.23 (ebook) | DDC 634.9/618—dc23
LC record available at https://lccn.loc.gov/2019036316
LC ebook record available at https://lccn.loc.gov/2019036317

Printed in the United States of America

HC 10 9 8 7 6 5 4 3 2 1
PB 10 9 8 7 6 5 4 3 2 1

Alissa Thielges, editor
Deb Miner, series designer
Ciara Beitlich, book designer
Aubrey Harper, photo researcher

Photos by Shutterstock/Alin Brotea cover, 8–9, 16; Dreamstime/Elantsev 1; iStock/buzzanimation 3; Shutterstock/wassiliy-architect 4–5; Shutterstock/xpixel 6–7; iStock/sswartz 10–11; Alamy/AB Forces News Collection 12–13; Alamy/USFS Photo 14–15